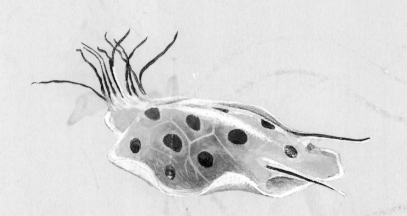

A Joshua Morris Book
Published by The Reader's Digest Association, Inc.
Copyright © 1992 Victoria House Publishing Ltd.
All rights reserved. Unauthorized reproduction,
in any manner, is prohibited.
Printed in Hong Kong.
Library of Congress Catalog Card Number: 92-60795
ISBN 0-89577-449-6
10 9 8 7 6 5 4 3

Reader's Digest Fund for the Blind is publisher of the Large-Type Edition of *Reader's Digest*. For subscription information about this magazine, please contact Reader's Digest Fund for the Blind, Inc., Dept. 250, Pleasantville, N.Y. 10570.

UNDERWATER

NATURE SEARCH

Consultants
ANDREW CLEAVE
DR. STEVEN WEBSTER

Illustrators
EVA MELHUISH
NEIL BULPITT
BRIN EDWARDS

a Joshua Morris book
from The Reader's Digest Association, Inc.

CONTENTS

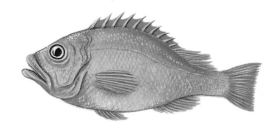

HOW IT WORKS

UNDERWATER

The oceans of our planet are deep, mysterious places, full of amazing animals and plants. Enter this underwater world of color and surprises, and discover the strange and wonderful creatures that live there. Some are hidden or camouflaged—use your magnifying glass to search the oceans and see what you can find. Look for this magnifying glass symbol. It indicates which creatures you should look for.

There are other challenges to test your powers of observation. For instance, you'll find an hippopotamus in a very unusual place!

There are answer pages at the back of the book as well as a glossary, where you can find out about the creatures that appear on every page.

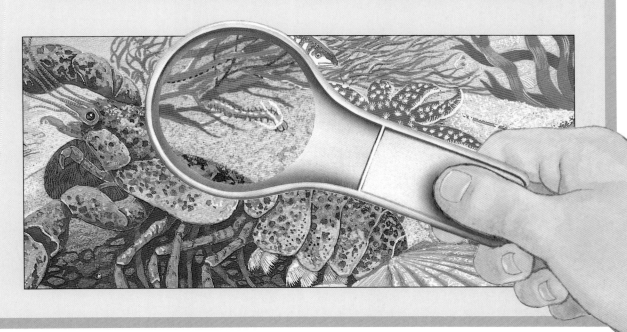

A WORLD APART

Far out in the Pacific Ocean, off the coast of South America, lie the Galapagos Islands — a forbidding group of volcanic islands once called the Enchanted Isles. The islands lie on the equator and cover an area of more than 3,000 square miles. Since the islands are warmed by the hot, tropical sun, creatures such as turtles, boobies, iguanas, and hammerhead sharks can survive here. But because the cliffs and rocks provide shelter from the hot sun and a cold sea current sweeps up from the Antarctic, animals usually found in colder climates are able to live on the islands and in the surrounding waters. The cold ocean current also carries nutrients that provide plenty of food for the creatures that make their homes on and around the Galapagos Islands. Some of these creatures never leave the islands, while others roam the oceans, returning to breed or seek shelter.

How many animals that fly can you find?

How many egg-laying animals can you find?

There are four Sally Lightfoot crabs and five red starfish in the picture. Can you find them all?

Some animals live on land, but catch their food in the sea. The sea lion is one. Can you guess which other animals do this?

LONG-DISTANCE FLIGHTS

The wandering albatross is one of the world's largest seabirds. Its great wingspan enables it to glide tirelessly across vast expanses of ocean in search of food. It may be away from land for over a year at a time.

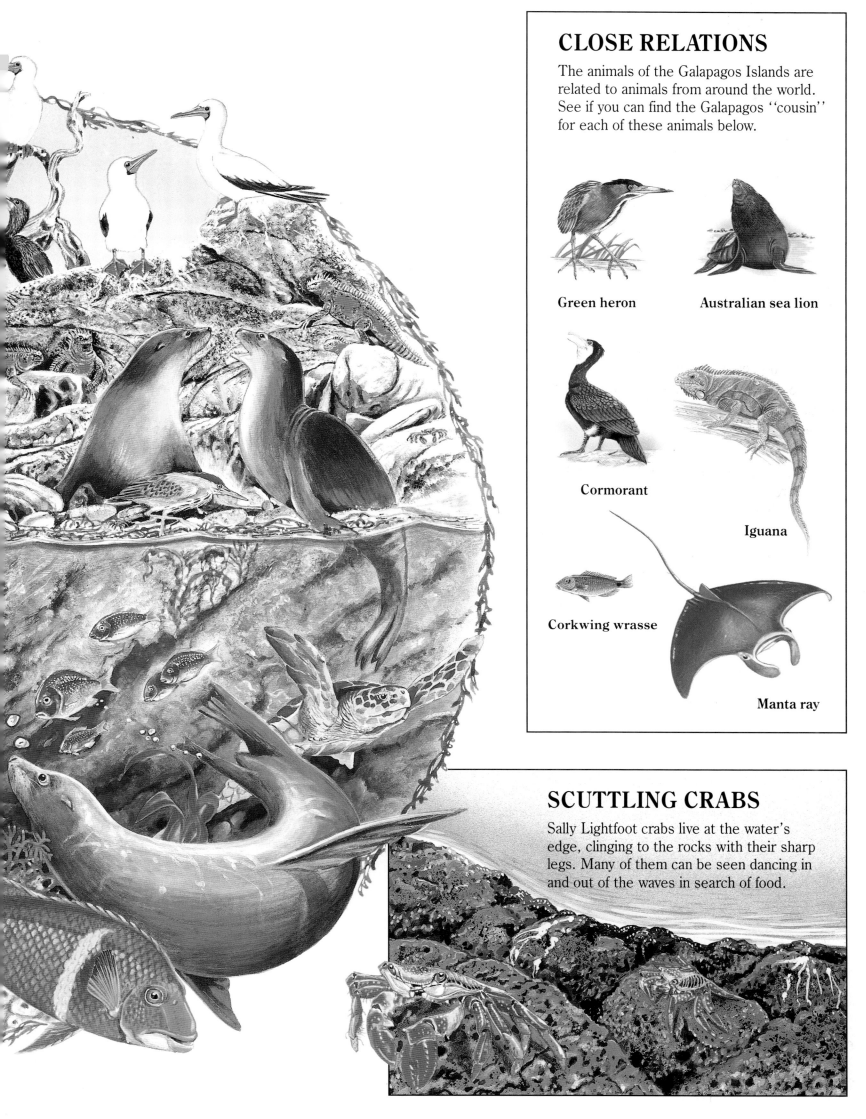

CLOSE RELATIONS

The animals of the Galapagos Islands are related to animals from around the world. See if you can find the Galapagos "cousin" for each of these animals below.

Green heron

Australian sea lion

Cormorant

Iguana

Corkwing wrasse

Manta ray

SCUTTLING CRABS

Sally Lightfoot crabs live at the water's edge, clinging to the rocks with their sharp legs. Many of them can be seen dancing in and out of the waves in search of food.

IN THE SHALLOWS

In shallow coastal waters, light streams down on the seabed below, encouraging plants to grow. Seaweed of various colors grows here: green seaweed grows near the shore, brown seaweed grows in the middle regions, and red seaweed is found in deeper water. Finding food and hiding from enemies keeps the sea creatures busy in these waters. While lobsters and crabs pick their way over stones and shells looking for a meal of dead fish, they must watch out for the large, silent octopuses ready to make a meal of them.

The variety of fish in these shallow waters includes shoals of tuna, mackerel, small sand eels, dogfish, and cod. Occasionally a humpback whale will venture into coastal waters to feed on these large schools of fish. Starfish and sea urchins, with their tough outer coverings, make these shallow waters home, as do soft-bodied bivalves, such as mussels, sheltered in their hinged shells, their siphons filtering water for food.

DANCING SCALLOPS

Most of the time scallops remain motionless on the seabed with their shells slightly open. If danger threatens, they snap their shells shut, forcing water out, which causes them to shoot backward. By repeating this action many times, they can swim through the water in a curious, jumping dance.

CAN YOU FIND THEM?

The sand and mud of the seabed is an ideal place for sea creatures to hide.

Worms are an important part of fishes' diets. For protection, worms hide in the mud with only their heads showing.

The horseshoe crab has a dull-colored shell shaped like a horseshoe. Its spiky tail stops large fish from eating it.

WHAT DO THEY EAT?

There are millions of microscopic plants, known as phytoplankton, in the sea. They are eaten by millions of tiny shrimp and sea snail larvae. In turn, the larvae are eaten by small sand eels. The sand eels are food for one of the largest creatures of all — the humpback whale. In one gulp, the whale can scoop up more than 10,000 sand eels!

Humpback whale

Sand eels

Larvae

Phytoplankton

The body of the pipefish has a tough armor-like covering. It swims almost upright.

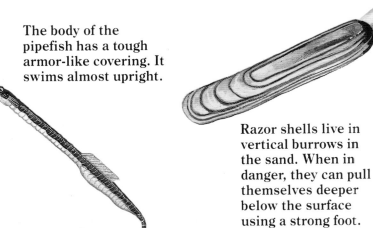

Razor shells live in vertical burrows in the sand. When in danger, they can pull themselves deeper below the surface using a strong foot.

Sand eels live in huge schools. Sometimes they burrow into the sand where they are very hard to find.

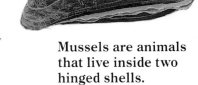

Mussels are animals that live inside two hinged shells.

A REEF OF COLORS

Running down the east coast of Australia is one of the wonders of the natural world. A vast coral reef has been formed by millions of living organisms. Coral is formed by tiny creatures called polyps, which build a "skeleton" of limestone around themselves. Large numbers of polyps join together to form a colony. As new polyps grow, the limestone formation grows. Over thousands of years, a huge wall of coral has risen from the ocean bed. Because of the warm water, the variety of life ensures that there is plenty of food and shelter for millions of sea animals.

The Great Barrier Reef is the only natural structure on this planet which can be seen from space — and it is in danger. A prickly-looking starfish, called the crown-of-thorns, feeds on coral. It turns its stomach outward, presses it against the coral, and absorbs the coral. When these starfish feed together, all that remains are large patches of dead, white coral.

SHARKS

Sharks cruise along the edge of the reef in search of food. The warm, clear water and the abundance of small fish attract many species of sharks. Some live on the seabed, partially hidden in the sand, some swim in and out of large crevices in the coral, and others swim in the open sea on the lookout for an easy meal.

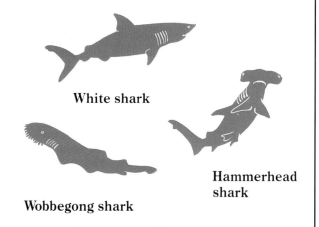

White shark

Wobbegong shark

Hammerhead shark

CAN YOU FIND THEM?

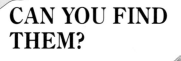

The squid has a soft, boneless body.

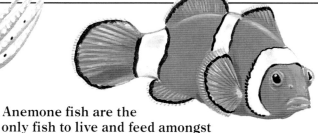

Anemone fish are the only fish to live and feed amongst the stinging tentacles of sea anemones.

Colorful sea slugs move over coral in search of food.

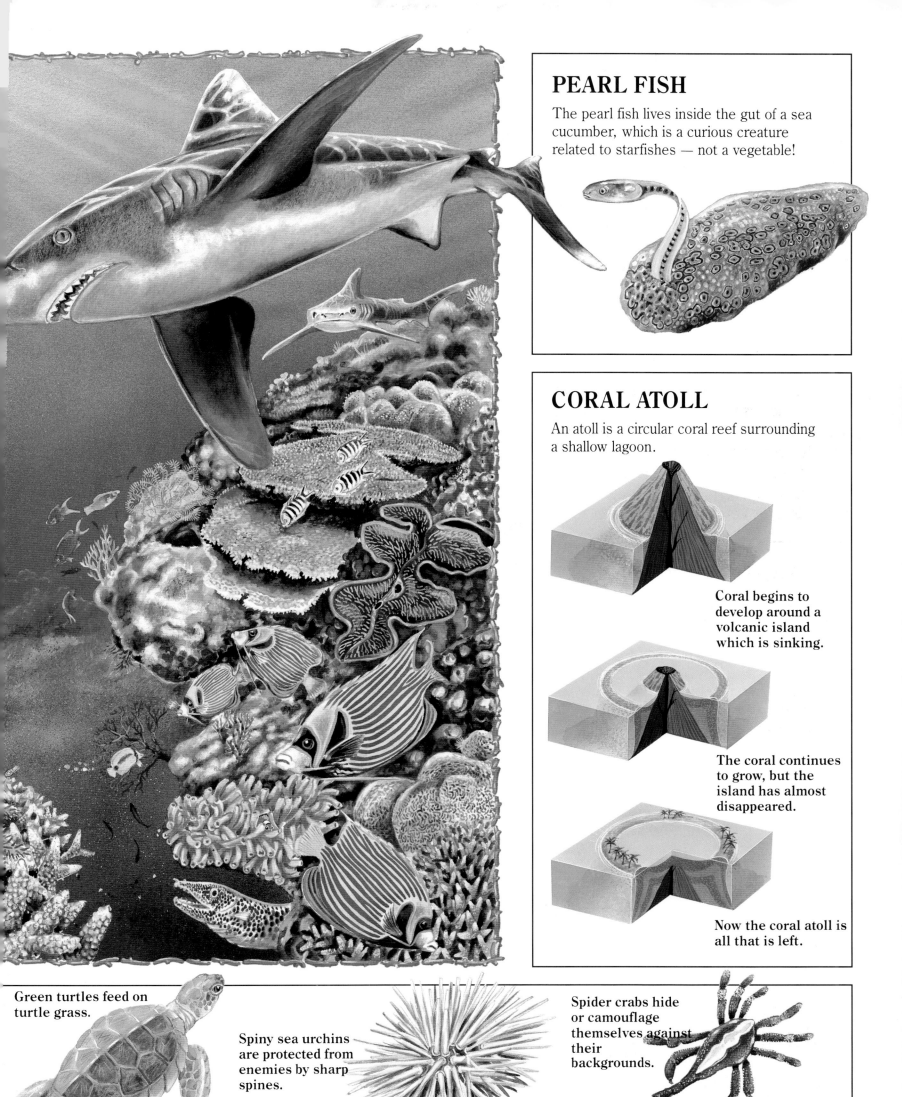

PEARL FISH

The pearl fish lives inside the gut of a sea cucumber, which is a curious creature related to starfishes — not a vegetable!

CORAL ATOLL

An atoll is a circular coral reef surrounding a shallow lagoon.

Coral begins to develop around a volcanic island which is sinking.

The coral continues to grow, but the island has almost disappeared.

Now the coral atoll is all that is left.

Green turtles feed on turtle grass.

Spiny sea urchins are protected from enemies by sharp spines.

Spider crabs hide or camouflage themselves against their backgrounds.

SARGASSO PUZZLE

The Sargasso Sea is approximately two million square miles of clear, deep blue water in the North Atlantic Ocean. Its name comes from the Portuguese word for seaweed — *sargaco*. Vast mats of seaweed grow here, attracting life above and below the surface. In this dense underwater forest, many of the fish have developed over millions of years to look like the Sargasso weed. They even behave like it, drifting and swaying slowly with the current. Jellyfish trail long tentacles through the open water; their stinging threads mean death to most small fish if they touch them. Tiny nomeid fish dart in and out of the tentacles of the Portuguese man-o-war, while giant oceanic sunfish make their stately way through open water. These fish can grow up to thirteen feet long!

When adult eels from the freshwater rivers of Europe and North America are ready to mate, they swim thousands of miles to the Sargasso Sea. Here they lay their eggs and then die. When old enough, the young eels make the long journey back to the freshwater homes of their parents until they too return to the Sargasso Sea to start the cycle again.

EEL ESCAPE!

This little eel is trying to escape from the sawfish.

Can you save him by finding the only route through the seaweed back to the other eels?

SEAWEED HOME

There are three Sargasso fish and three Sargasso crabs hidden in the picture.

Can you find them?

15

HARBOR LIFE

Harbors, whether naturally occurring or manmade, are sheltered places where boats are protected from the sea. In this harbor the outer seawall shelters the rest of the harbor where many plants and animals live. Here they are safe from the worst storms and the strongest waves. Delicate creatures like sea anemones and tiny fish can feed in calm water inside the harbor, while rough seas rage further out.

Many other plants and animals live in harbors. Some are only visitors, looking for food or shelter, while others live here all the time. The harbor seal and porpoise swim in from the open sea hunting for fish. On the harbor bed, small fish feed on worms, crabs, and other small creatures. The crabs scuttle among the seaweed and sponges scavenging for scraps of food. Wooden pilings, where boats are tied up, are encrusted with goose barnacles, and are often riddled with holes made by shipworms.

SEA POLLUTION

Sea pollution is caused mainly by oil and chemical leaks from ships and from human sewage. Poisonous chemicals dissolve in the sea and shellfish and other sea creatures absorb them. When people eat tainted seafood they can fall seriously ill. Sea animals also suffer; birds drown when their wings become clogged with oil and seals also die from chemical poisoning.

CAN YOU FIND THEM?

The harbor provides protection for the animals that live there. Small fish can feed safely among seaweed which grows on the seabed and the harbor walls. Stones, wooden stumps, anchors, and even garbage provide safety.

Hermit crabs have soft bodies, so they hide themselves in empty shells, coming halfway out to feed and walk around.

The gray sea slug eats beadlet anemones, whose stinging cells do not harm it.

The shipworm is a clam with a long, worm-like body which burrows into wood to hide itself.

The two-spotted goby swims near the bottom in search of food. Its color provides good camouflage among the weeds.

The porcelain crab uses its long claws to scoop mud and sand toward its mouth, where it filters out tiny bits of food.

The strange markings and unusual body shape of the pogge make it hard to find as it lies still on the seabed.

ODDITIES AT IZU

The seas that lap the rocky cliffs of Japan's Izu Peninsula hide a surprisingly colorful world, known only to a few deep-sea divers and the mysterious creatures that live there. The murky surface waters of the sea, often whipped into large, heavy waves by winter storms, prevent any light from reaching the depths below.

Human divers need to take powerful underwater lights with them to illuminate the richly-colored corals, brightly-striped moray eels, starfish, urchins, and jellyfish. These creatures all compete with each other for the most outstanding color schemes. These deep waters are too cold to sustain the tropical reef-building corals, but many large sea fans, soft corals, and sponges thrive here. They live off the rich food supplies carried by the strong ocean currents sweeping past the rock faces. Swimming among the currents in a constant search for food are hunters like the squid and moray eel. Smaller fish, such as gobies, flit up and down among the coral branches, searching for scraps and then hiding in the crevices for safety. Even fixed feeders like sea anemones are able to catch plankton and small fish which drift by in the sea's currents.

These waters contain some odd plants and animals. Can you find those that really don't belong?

A GIANT OF A CRAB

The giant Japanese spider crab's long jointed legs span over six feet! Pincers at the end of its front legs can grip tiny pieces of food, while the spiny ends of its walking legs enable it to move over coral.

CORALS

Corals are colorful animals which feed on very tiny food particles. They have many tiny mouths surrounded by waving tentacles. These tentacles trap small pieces of food swept past them by the sea current. Fan corals form branching colonies and brain corals form large cushion shapes.

THE OCEAN DEEP

The ocean depths are cold and dark, yet many creatures are able to survive here. The fierce-looking angler fish with its huge, gaping mouth waits silently for small fish to swim by, ready to catch them in one gulp. The conger eel lurks in the dark, waiting for careless fish to venture too close, while the gulper eel uses its enormous mouth to swallow prey which can be larger than the gulper eel itself! The giant squid also lives in the icy ocean depths, gliding rapidly through the water in search of food. The squid has only one real enemy – the enormous sperm whale which can dive deep into the ocean to hunt the squid. Fierce battles are fought far below the surface by these giants of the deep, but the whale, with its powerful jaws and sharp teeth, is usually the winner.

One of the animals in the picture is a mammal, not a fish. It uses lungs to breathe and nurses its young. Do you know which one it is?

CHALK CLIFF ANIMALS

Foraminiferans are very tiny creatures which live in delicate, chalky shells. They live near the surface of the sea, drifting with the current, but once they die they fall slowly to the ocean floor. Billions of them fall to the seabed, and their chalky shells form thick layers, eventually turning to chalk rocks.

CAN YOU FIND THEM?

Remaining hidden is very important, even in the darkness of the ocean depths. Some creatures lie flat on the bottom, some crawl into holes, and some have confusing shapes or colors. How many of these can you find?

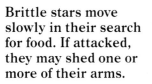

The grenadier's large eyes help it find its favorite food — the tiny shrimp which feed on debris on the seabed.

Brittle stars move slowly in their search for food. If attacked, they may shed one or more of their arms.

FISH THAT CAN SHINE

In the depths, all light comes from creatures that live there. Some deep-sea fish emit colored light to attract prey, some flash to startle an enemy, and some attract a mate by glowing in the dark.

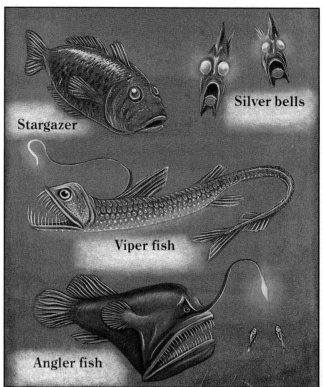

Stargazer

Silver bells

Viper fish

Angler fish

The small delicate shrimp is related to crabs and lobsters.

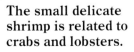

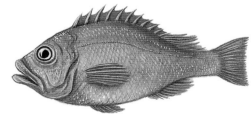

Redfish feed on small fish, shrimp, and shellfish.

Flatfish, like the plaice, lie still on the seabed. Eyes on top of the head help them see their prey approaching.

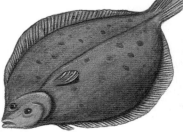

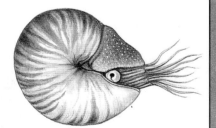

Most of the nautilus's soft body is protected by a coiled shell. It uses its powerful tentacles to grasp food and push the food into its mouth.

PUZZLE ANSWERS

Did you find all the hidden animals with your magnifying glass?
In the following pictures, the position of each hidden animal is
indicated by a circle. Answers to the questions are given as well.

A WORLD APART

The flying animals are the 1 waved albatross, 3 red-footed boobies, and 1 lava
heron. (The flightless cormorant does not fly.)

There are 35 egg-laying animals in the picture. Rays, most sharks, and sea
lions do not lay eggs, but all the other animals do!

The sea lion, flightless cormorant, penguin, marine iguana, red-footed booby,
and lava heron live mainly on land and get their food from the sea.

All these animals are related:

The green heron and the lava heron.
The common cormorant and the flightless cormorant.
The Arctic sea lion and the Galapagos sea lion.
The South American iguana and the marine iguana.
The manta ray and the spotted ray.
The corkwing wrasse and the Galapagos wrasse.

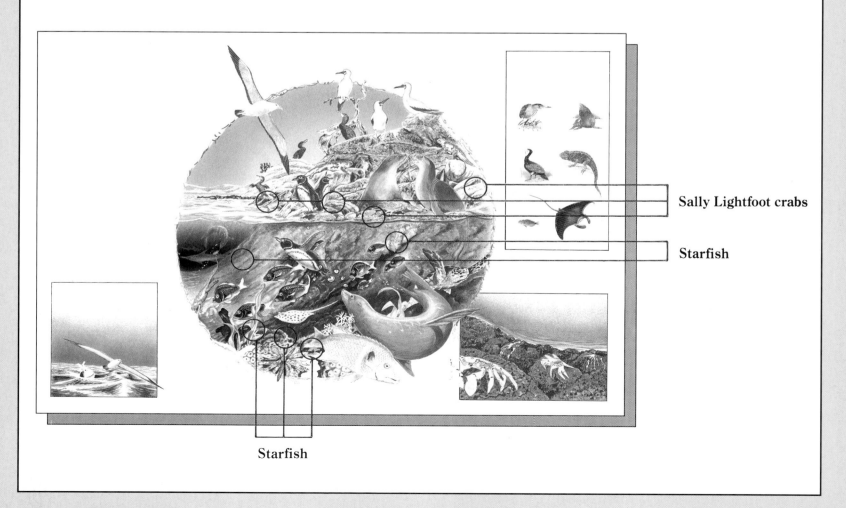

Sally Lightfoot crabs

Starfish

Starfish

IN THE SHALLOWS

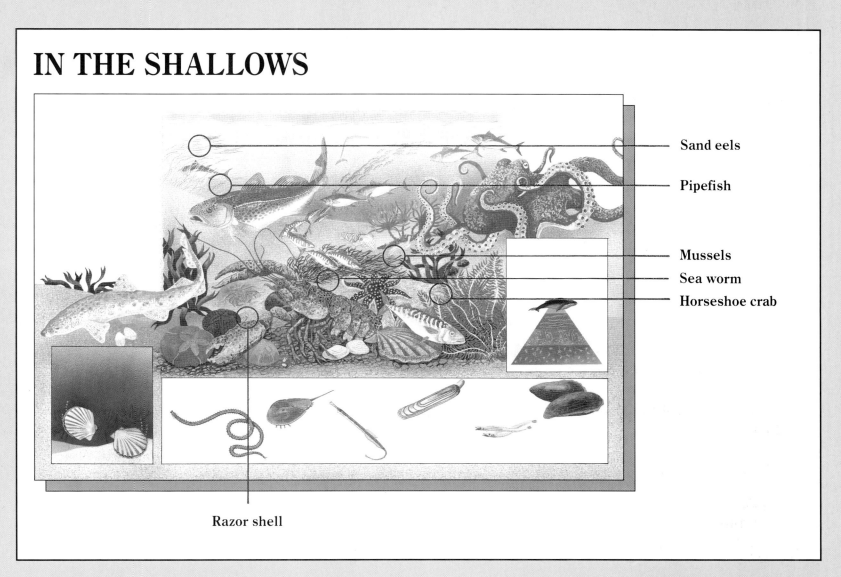

Sand eels

Pipefish

Mussels

Sea worm

Horseshoe crab

Razor shell

A REEF OF COLORS

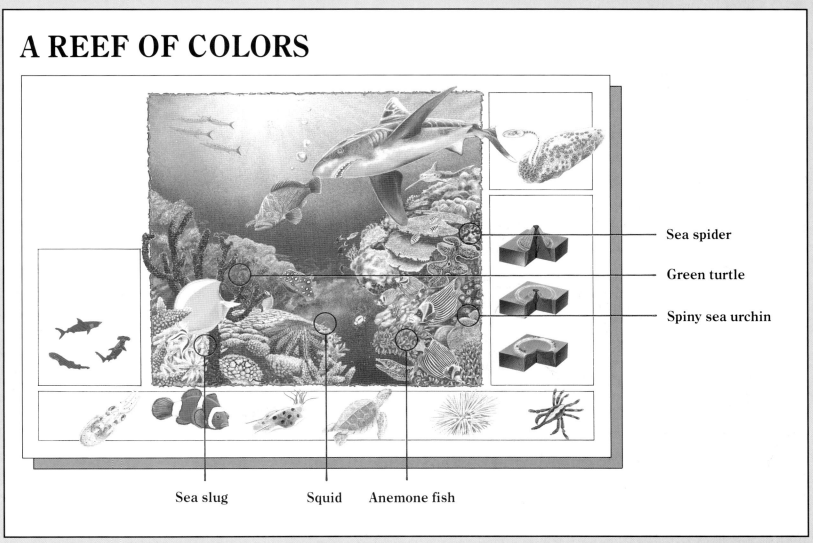

Sea spider

Green turtle

Spiny sea urchin

Sea slug   Squid   Anemone fish

SARGASSO PUZZLE

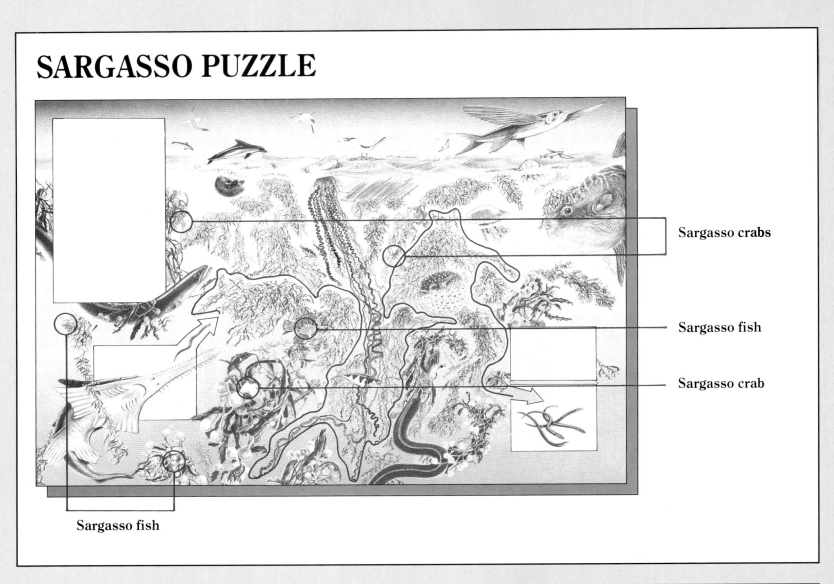

Sargasso crabs

Sargasso fish

Sargasso crab

Sargasso fish

HARBOR LIFE

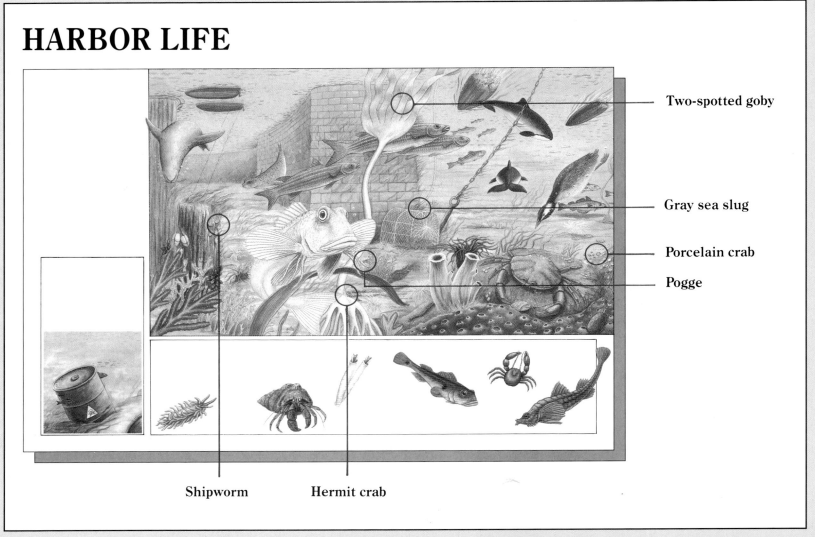

Two-spotted goby

Gray sea slug

Porcelain crab

Pogge

Shipworm   Hermit crab

ODDITIES AT IZU

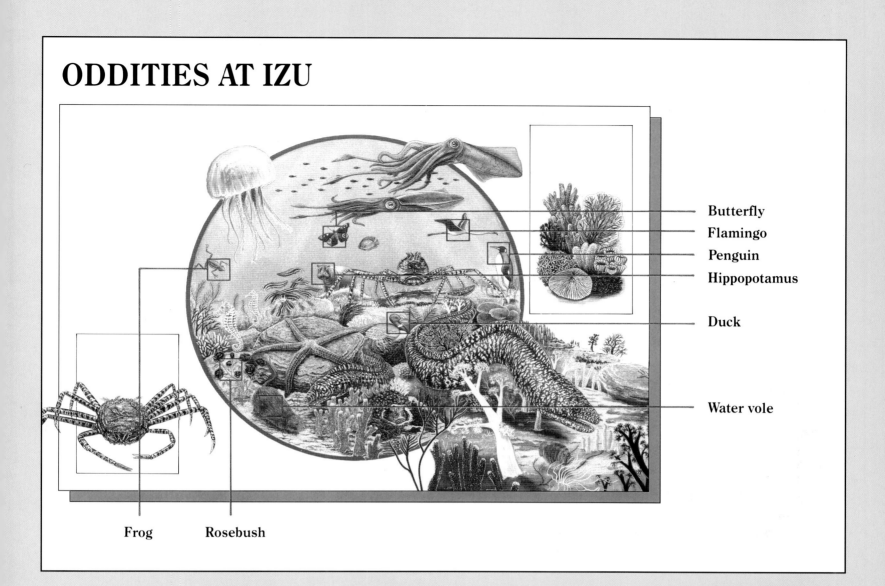

Butterfly

Flamingo

Penguin

Hippopotamus

Duck

Water vole

Frog   Rosebush

THE OCEAN DEEP

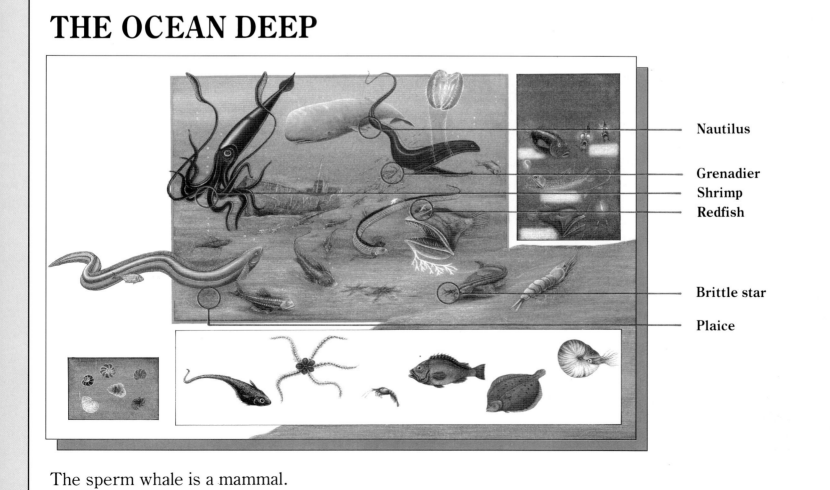

Nautilus

Grenadier

Shrimp

Redfish

Brittle star

Plaice

The sperm whale is a mammal.

GLOSSARY

A WORLD APART
pages 8–9

Australian sea lion
The male sea lion fights for its territory. It may have up to 40 females to look after.

Corkwing wrasse
The corkwing wrasse grows to about 6 inches in length. Its color may vary from green to reddish-brown.

Cormorant
The cormorant is a powerful bird which swims through water searching for fish. It holds up its wings to dry them.

Flightless cormorant
The flightless cormorant is found only on the Galapagos Islands. It cannot fly.

Green heron
The green heron has short legs and feeds mainly at night, diving into the water for its prey.

Green turtle
The green turtle lives in warm, coastal seas and eats mainly sea plants.

Hammerhead shark
The hammerhead shark has a flattened head, with eyes and nostrils set wide apart.

Iguana
Iguanas are lizards which live mostly in trees and eat insects. They live mainly in tropical America.

Lava heron
The lava heron has a narrow head. It wades silently along the shore, looking for small fish and crabs to eat.

Manta ray
The manta ray is an enormous sea fish with a flattened body. It has two fins on its head.

Marine iguana
The marine iguana lives on the coasts of the Galapagos Islands. It can swim and dive well to find seaweed.

Penguin
Penguins live only in the southern hemisphere. They cannot fly but can swim well.

Red-footed booby
Red-footed boobies are large seabirds with sharp bills. They catch fish by diving into the water from the air.

Sally Lightfoot crab
These crabs scuttle up and down the rocky shore, scooping up tiny morsels of food left by the waves.

Sea lion
The sea lion is a mammal which has flippers instead of arms and legs. It is a powerful swimmer and can move on its flippers when on land.

Silver-sided grunt
Silver-sided grunts live in shoals for safety from sea lions. The sea lions hunt the shoals and try to break them up.

Spotted ray
The spotted ray is a sea fish with a flattened body. It is related to sharks and eats crabs, mollusks, and fish.

Starfish
Starfish usually have 5 arms, but some can have over 40. They are related to sea urchins.

Wandering albatross
The wandering albatross has the greatest wingspan of any living bird — up to 11.5 feet.

Waved albatross
The waved albatross has long, narrow wings. It feeds on squid and other sea creatures.

Wrasse
The wrasse has teeth which it uses for eating other fish and shell-living sea creatures.

IN THE SHALLOWS
pages 10–11

Cod
The cod is a large fish which feeds near the seabed. It eats fish, crabs, sea urchins, and shellfish.

Cuttlefish
The cuttlefish is a type of mollusk. It has a chalky inside shell, 8 arms and 2 tentacles.

Dogfish
The dogfish belongs to the shark family. Its skin is rough and spotted.

Flounder
The flounder is a flatfish which lives camouflaged on muddy seabeds.

Heart urchin
The heart urchin usually lives buried in sand. It feeds on tiny scraps of waste matter.

Horseshoe crab
The horseshoe crab is related to spiders, and is not a crab at all. It has a horseshoe-shaped shell and a long, pointed tail.

Humpback whale
The humpback whale is one of the largest whales. Its huge mouth can swallow whole shoals of fish.

Larvae
Larvae are the young stages of a variety of sea creatures, such as crabs, mollusks, and fish.

Lobster
The lobster feeds on scraps of food on the seabed, and defends itself with its large pincers.

Mackerel
The fast-swimming mackerel lives in large shoals near the surface.

Mollusk
A mollusk is a soft-bodied creature which lives inside a shell.

Mussel
The mussel is a mollusk with two hinged shells. It attaches itself to rocks with tough, sticky threads.

Octopus
The octopus is a mollusk with no shell. It has 8 tentacles and a tough beak.

Pipefish
The pipefish has a long, thin body. It sucks in small creatures through its thin mouth.

Plankton
Plankton are microscopic drifting plants and animals which live near the surface of the sea.

Razor shell
This animal has two hinged shells. Its foot sticks out of one end and it breathes through the other.

Sand eel
The sand eel feeds during the day and spends the night hidden in sand.

Scallop
The scallop is a mollusk with a two-part shell — the top is flat and the bottom is saucer-shaped.

Segmented worm
The body of the segmented worm is divided into many tiny sections joined end to end.

Spider crab
The spider crab has long legs and a rounded body. Its shell is often covered with seaweed.

Starfish
Starfish move by using tiny tubes and suckers on the underside of the body.

Tuna
The tuna is a powerful and fast-swimming fish. It grows to over 6 feet in length.

A REEF OF COLORS
pages 12–13

Anemone fish
The anemone fish lives safely among the stinging tentacles of the sea anemone.

Barracuda
The barracuda is a long fish with large, sharp teeth. It can grow to almost 6 feet long.

Blue shark
All sharks have very keen senses. They can smell and taste the water to tell whether prey is near.

Box fish
The box fish is protected by a covering of shell-like armor.

Crown of thorns starfish
The long spines on its body give this starfish its name. It eats coral, causing damage to reefs.

Emperor angelfish
When young, the markings of this fish are blue and white circles.

Green turtle
The green turtle is protected by its thick, bony covering. The females lay their eggs on land.

Grouper
The grouper is a heavy fish. Most groupers begin adult life as females, and become males as they get older.

Hammerhead shark
The hammerhead shark has its eyes and nostrils on each side of the hammer-shaped head.

Jack
The stripes of the jack break up its outline, making it difficult for predators to see it.

Jellyfish
The jellyfish has no skeleton. It has a jelly-like body between two layers of cells.

Pearl fish
Pearl fish live inside the bodies of animals such as sea urchins and sea cucumbers.

Sea slug
Sea slugs are mollusks with little or no shell.

Speckled moray eel
This tropical eel lives among rocks and corals.

Spider crab
This slow-moving spider crab is camouflaged against a red coral background.

Spiny sea urchin
The spiny sea urchin is found in shallow parts of the coral reef.

Squid
Most squid grow no longer than 19 inches. They have 10 arms.

Squirrelfish
Squirrelfish are usually a reddish color. They are active at night and have large eyes.

Surgeon fish
Surgeon fish use small, sharp teeth to scrape plants and animals off coral and rocks.

White shark
The white shark grows to over 23 feet long. It preys on fish, sea lions, and other sharks.

Wobbegong shark
The mottled markings and flat shape of the wobbegong shark camouflage it against the seabed.

SARGASSO PUZZLE
pages 14–15

By-the-wind-sailor
This animal has a sail-like float which catches the wind, helping it travel across water.

Flying fish
Flying fish do not fly — they leap from the water and glide, using their fins.

Freshwater eels
These eels live in rivers and ponds. They travel to the Sargasso Sea to lay their eggs.

Jellyfish
The jellyfish is related to the sea anemone. It has stinging tentacles.

Nomeid fish
The nomeid fish lives among the tentacles of the Portuguese man-o-war without being stung.

Oceanic sunfish
This spade-shaped fish can weigh as much as 2,200 pounds. It eats animals like jellyfish.

Portuguese man-o-war
The Portuguese man-o-war is made up of a float and long, stinging tentacles.

Puffer fish
The puffer fish defends itself by puffing up its body, showing long, sharp spines.

Sargassum crab
The Sargassum crab is well camouflaged to blend in with seaweed.

Sargassum fish
The Sargassum fish lives among floating masses of seaweed. It crawls through the clumps of weed.

Sargassum trigger fish
The Sargassum trigger fish has a long spine on its back fin.

Sawfish
The sawfish uses its flattened, toothed snout to dig out small creatures from the seabed.

Tropic bird
The tropic bird flies low over the sea, hunting for fish and squid.

Violet sea snail
The violet sea snail floats on a raft of bubbles at the sea's surface.

White-sided dolphin
This sea mammal has a torpedo-shaped body and a tall, curved fin on its back.

HARBOR LIFE
pages 16–17

Barnacle
The barnacle is related to crabs and lobsters. It cements its head to rock, wood, or other shells.

Common porpoise
This porpoise is a sea mammal which spends its whole life in the water. It may live for 12 or 13 years.

Common seal
The common seal is a sea mammal with a sleek body and paddle-like flippers.

Dog whelk
The dog whelk breathes through a long tube or siphon. It eats mussels.

Goose barnacle
The goose barnacle gets its name from the shape of the stalk by which it hangs on to a rock.

Gray sea slug
The gray sea slug feeds on sea anemones by eating their tentacles. Somehow, it avoids being stung.

Hermit crab
The hermit crab has a soft body and lives in an empty sea shell.

Lemon sea slug
The lemon sea slug lives under stones on rocky shores and feeds on sea sponge.

Mullet
The mullet lives around the shore and in estuaries. It eats algae and small creatures living in the mud.

Mussel
Mussels are often found in large groups attached to each other by threads.

Pogge
The pogge is a small fish with a long body and tapering tail.

Pollack
The pollack is related to the cod. Small fish are found inshore but, as they grow, they move to deep waters.

Porcelain crab
This small crab is related to squat lobsters and hermit crabs. It lives under stones.

Ragworm
The ragworm lives in a burrow in mud or muddy sand. Some ragworms grow up to 19 inches long.

Red gurnard
The red gurnard has fins with "feelers." The fish uses them to find food.

Sea anemone
The sea anemone has tube-like bases which are attached to a rock. The mouth is surrounded by stinging tentacles.

Sea hare
The sea hare is a slug-like sea creature that feeds mainly on sea lettuce.

Sea squirt
The sea squirt looks like a bag of jelly. It gets food from water pumped through its body.

Shipworm
The shipworm uses its shell to bore into wood. It lines the holes with a chalky substance.

Shore crab
The shore crab grows up to 3 inches across. It looks for food at night and at high tide.

Slavonian grebe
This grebe feeds mainly on insects and fishes. It builds a floating nest attached to plants.

Two-spotted goby
This goby grows to about 2.4 inches long. It lives in shallow seas near the coast.

ODDITIES AT IZU
pages 18–19

Butterfly
A butterfly is an insect with club-shaped "feelers" and broad wings.

Butterfly fish
Butterfly fish are usually found on coral reefs.

Catfish
The young of certain catfish may gather into a "feeding ball" of several thousand.

Duck
A duck has webbed feet and its feathers are waterproof.

Fin reef squid
The squid is a mollusk. It has 10 arms and swims very fast.

Flamingo
The flamingo feeds by holding its bill underwater, to filter out small particles of food.

Fringe-headed blenny
These fish have tufts of skin which look like a fringe above the head.

Frog
A frog is an amphibian, which means that it can live on land or in water.

Giant Japanese spider crab
The leg span of some of these giants can reach 26 feet!

Hippopotamus
The hippopotamus spends much of the day in water and comes out at night to graze.

Penguin
The penguin cannot fly, but its webbed feet and flipper-like wings mean that it swims well.

Sea horse
The sea horse is a small fish with a tube-like snout. It swims upright.

Speckled moray eel
This eel lives in holes and crevices in rocks in tropical oceans.

Starfish
Starfish are related to sea urchins. They eat sea creatures such as sponge or coral.

Stinging jellyfish
The jellyfish is related to the sea anemone. It has stinging tentacles.

Striped morwong
The striped morwong is often accompanied by small blue cleaner wrasses which keep it clean.

Tiger moray eel
This brightly-colored eel uses its teeth for grinding up crabs.

Water vole
The water vole lives close to water and swims by paddling quickly.

THE OCEAN DEEP
pages 20–21

Arborescent deep-sea angler fish
This fish has light organs above and below its mouth. Certain muscles control these light organs.

Brittle star
The slender arms of the brittle star snap off easily, but new arms soon grow to replace them.

Chimera
The chimera, or rabbit fish, has no scales, and its tail tapers to a long thread.

Conger eel
The large conger eel lurks in dark crevices, ready to seize its prey of fish, squid, crab, or lobster.

Deep-sea angler fish
The angler fish uses a dangling light organ to tempt smaller fish close to its enormous mouth.

Foraminiferan
A foraminiferan is a tiny creature which lives inside a chalky shell. White chalky cliffs are made of foraminifera.

Giant squid
The giant squid is a mollusk with 10 arms.

Grenadier
The grenadier's large eyes help it to find prawns just above the seabed.

Gulper eel
The gulper eel is almost all head and mouth. Its tiny eyes are perched on its snout.

Herring
Herrings are found in deep ocean waters, but they lay their eggs in shallow coastal waters.

Jellyfish
The jellyfish has a transparent body which is made up mainly of water. Its tentacles are covered with stinging cells.

Lantern fish
The lantern fish lives deep in the ocean, but rises to the surface at night. It has tiny light spots on its side.

Monkfish
The monkfish has a flattened body and is related to sharks. It lies buried in sand, waiting for its prey.

Nautilus
Most of the nautilus's shell is empty and helps the nautilus to float at any depth.

Oarfish
The oarfish has a very long, flattened body with a small head and fringe-like fins.

Plaice
The plaice has a flattened body. As a larva, its body changes, so that both eyes are on the same side of the head.

Prawn
The prawn scavenges near the seabed and can quickly dart out of the way of hungry predators.

Redfish
The redfish lives near the seabed and feeds on small fish, shrimps, and shellfish.

Shrimp
The shrimp has an almost transparent body which hides it from its predators.

Silver bell
The silver bell has rounded, bulging eyes that point upward. It can see prey outlined against light from above.

Sperm whale
The sperm whale is the largest toothed whale. It grows to almost 66 feet long.

Stargazer
The stargazer has light organs behind its eyes and poisonous glands on its body.

Tripod fish
The tripod fish has fins with long spines which help it to rest on the seabed.

Viperfish
The viperfish has very long teeth and a huge mouth. It has flashing lights on its underside.